EARTH TO MARS

Humanity's Interplanetary Journey

Dr. Friday E. Abu

*This book is dedicated to all those who are
positively curious about the possibility of anyone's
departure from the Earth, go to, live well and
thrive greatly on the red planet, Mars.*

CONTENTS

PREFACE

Humanity's state of scientific and technological advancement has every requisite knowledge, brilliance, intelligence and wisdom packs to prepare for manned crew interplanetary journey to the red planet, Mars. Yes, the state of the world's present breakthroughs, in rocket science, is already receiving the practically perfecting touches for the take-off for Mars. It is a welcomed stage to really test man's capacity to dream of further future longer cosmic expeditions like going to other star systems, places outside our own solar system. For man to be able to successfully land on Mars, live, walk and work on its surface, is to pass a great test thrown at us by our very own solar system. An aftermath of which would be the necessary courage to further the quest.

Mars is the closest among the planets which humanity would find easiest to fine-tune for man's habitation. [Please, take note of that clause, 'easiest to fine-tune' because Mars is not the closest to the Earth.] To put differently, Mars would pose the least challenges to man among the options availably considered for colonization in our solar system. Be it as it may, the challenges placed by the red planet before the world scientific and technological bodies are enormous but, the various ways of conquering them have been discerned. If Mars can be colonized, any other planet, moon and cosmic rock out there would be far easily handled than if man does not pass the colonization test from this 'test tube' called Mars.

Man is not caged and locked-up on Earth. Man is meant to become multi-planetary and as a matter of fact, colonize planets of other star systems. Humanity must understand this; the universe does not always present everything rosy. Electricity was discovered to dispel darkness and so, something – may be a way or various ways – must be discovered to disperse the inhabitability of the red planet in question. This is my point; man surmounted various obstacles via discoveries and so is Mars a cosmic obstacle which must be conquered via currently attained technological and rocketry heights. That is the beginning of man's becoming a multi-planetary species and a great leaping step to think beyond that. If done, you and I will have at least two options placed before us – Earth or Mars – to live on.

Landing alone on the red planet is not the only thing

possible to humanity now but, also the well-doing of any human colony; both attainable via the world's stand, technologically. From reliable sources, I have gathered that those humans going to live on the red planet may do glaringly better in welfare than those on Earth.

Earth to Mars seeks to let us know that the red planet can be really made habitable for humans. That is not all; this book will make us understand how Mars can and will be made habitable. Again, some of the already taken steps to see to the habitability of this planet of our concern will be looked at. Also, this book will reveal some of the prospects available on the red planet. Then, Earth to Mars will look at the forecast welfare of the intending Martians as against that of the Earthlings and let's see if the colony on Mars will do better.

If indeed the universe has placed before humanity the test of making Mars habitable, then, let's wait to see how this test would be done and dusted by the scientifically technological minds of the present day, as will be shown in this book.

INTRODUCTION

Mars, the fourth planet from the Sun has always fascinated people with its reddish hue. Due to this hue, before the name 'Mars' was finally settled upon, the planet was called 'Mars Har Decher' meaning the Red One.

It was the Romans who later gave the name that has stuck, naming the red planet after their god of war and an agricultural guardian, Mars. Drawn by many who went into various researches and these people came-up with a resolute conclusion; it is scientifically and practically easier for humanity to colonize and live on Mars, than to think of achieving that on any other planet in our solar system.

One day, I asked a university professor this question, "Prof. which will be easier to terraform and live-on; Venus or Mars?"

He looked at me, smiled and said, "I'll put the answer in a question form." This professor looked at the blank papers on his desk, as though for the next move, for about four seconds and he asked, "Young man, which will you find easier to handle with your bare hands; a handle-less frying pan already frying on fire or an ice block?"

I instantly replied, "An ice block."

He then said, "You've rightly said Mars will be easier and that's the frank truth."

I had the answer above long before I asked but, I wanted to hear him buttress some facts, which he did. He (the professor), even added that Mars will be the cheapest for man to colonize than any other planet, at any given point in time.

At this moment, the red planet has no infectious organism or pathogens known to be alive yet. By this singular quality, it is definitely a place that should receive quick consideration from those persons who want to be found in an environment free of all disease-causing organisms. This is a great plus for the planet, which assuredly gives a vivid health-boost to all those thinking of making the red planet their permanent home.

The human colony on Mars will be self-sustaining; men and women fending for themselves and doing everything, while making the red planet become uniquely habitable, in some ways. Before this autonomy, Blue Origin, a rocket company single-handedly founded by the greatest online entrepreneur, Jeff Bezos and SpaceX, founded by Elon Musk and Tom Mueller will provide services unto the humans who landed on Mars. Other transporters are joining them in the future but, those two are ahead in the space race and may remain major suppliers, supplying needs to the human colony on Mars, when the time comes. Those two notable private companies will help transport humans alongside their belongings and needs to the red planet, to even start the colonization in the first place. Government agencies like the National Aeronautics and Space Administration (NASA) and European Space Agency (ESA) are in the space race but, their intention to deliver supplies to the future colony on Mars has not been clarified. The chain of supplies to the newly settled Martians remains unbroken until the human colony becomes self-sustaining.

The voyage to Mars on spaceship wouldn't in any way be stressful at all. From the interior systems being planned by the various intending transporters to the red planet, the manned voyage to Mars would be the 1st longest jolly ride in the history of humanity. It would be fun travelling to the red planet and the 6 – 7 months voyage may whisk-by time unnoticed. Time will not seemingly fly because of whisky but, because of beehive of refreshing recreational activities on board of the spaceships transporting humans to Mars. More of this comes ahead, in Chapter 6.

The red planet presents myriads of opportunities, if looked at keenly. Mars may be physically dry for now but, so wetted with riches, if only you can see it. Companies are going to be springing-up there and all sorts of skilled and non-skilled workers would definitely be needed. Wherever you're, if what you will be doing when Mars is ready for human colonization doesn't pay you as you want, someone may offer to transport you to Mars, give you a job and then, you gradually pay your voyage fee from some cool monthly pay. The question is; are you ready to become a Martian? Some individuals may be targeting the natural resources on this planet and are heading there. One of them may have rightly said, "There will be a lot of jobs on Mars!"

When men and women arrive on Mars, it would become a savored place of interesting mixture of diverse beautiful people. It would not be boring. Diverse cultural backgrounds from America, Africa, Asia, Europe etc. are already warming-up for the red planet. Believe this; I know wonderful people from the listed continents above who wish Mars is ready for colonization as we speak. Sweet men and women preparing to leave the Earth and relocate to Mars are already available. A particular couple I met said, "We just want to leave this polluted and pathogens-filled Earth, and try Mars when it's ready for the human colony." They are already saving towards the trip to the red planet. By the virtue of the personality of these people and others who are interested in being on this beautiful red planet, Mars will become a home to multi-integrated, diverse-cultured and happy people with a convergent future civilization.

CHAPTER ONE

A Mission Impossible?

Let's take a look again at our level of improvement and breakthroughs in rocket science on Earth before arriving at the answer to the question; is going to Mars an impossible quest? In 1903, the Wright brothers ended rightly in conquering gravity by producing the first authentically documented flying plane. From that year (1903) till now, gravity has been mastered-over by various improvements brought into flying machines until lift-off from the ground is no matter again for present day technological community. If you're given the specifications of the Apollo 11 Guidance Computer, which helped in landing men on the moon in 1969, you'll marvel that humanity achieved success using a 4KB RAM and 32KB Hard drive computer, back then. As we speak, if you're asked to go on any mission that such a computer will guide your spaceship, will you go? I bet you'll say, "No way! I'm not going on that suicide mission." But, that computer guided the astronauts over 50 years ago and humanity landed safely on the moon. From then until now, geometric developmental breakthroughs have been recorded in the area of computing, that we now talk about Terabytes (TB) and not Kilobytes (KB) anymore. Now, let's marry these two; conquering of gravity and guiding the traveling of a craft through space. The Wright brothers' historical breakthrough has been greatly improved upon that gravity conquering is far and exceptionally better than what it was in 1903. Overcoming gravity is now well improved upon that humanity is thinking of lifting about 450 tons to Mars, while carrying 1,000 passengers on board. Talking about guidance computer, the speed of information processing, data storage, analysis and decision-making of the present day guidance computer is about 99% better than what was used to land on the moon surface in 1969. Considering these two factors combined – which are still being improved upon, checked and re-checked – the lives and properties of men and women intending to go to Mars are very well preserved.

Another factor being put into great cognizance and worked indefatigably upon is the space suit. The suits used by the astronauts before now will not protect man as supposedly and therefore, improvement in suit design, suit protective capacity and suit flexibility have been thoroughly done to withstand whatever the Martian environment will present. The thin atmosphere of Mars, coupled with the non-available magnetic shielding has compelled the spacesuit designers and manufactures to consciously adopt an excellently-designed spacesuit that will absorb whatever is going to threaten life on this planet to be inhabited by man. It is known that Mars is cold and the temperature could fall to as low as –125 degrees Centigrade on a winter night, at the poles. But, you've got this all covered; whether you are out there on the Martian soil or in your Martian building. The radiation-shielding, temperature and pressure automated regulation capabilities of these suits are daily being reviewed and improved-on. From trustworthy sources so far, you will not need the spacesuit once you're indoor on Mars; every form of building and habitats planned for, will definitely incorporate those rightful protective devices and capabilities for you to be able do without the suit. Indeed, man will be well shielded on Mars.

Rovers are on Mars collecting data, processing and analyzing these data. From the collected data so far, man has been able to pinpoint exactly, with vivid precision, where the first human colony should land. This is to prevent any unnecessary rigmaroles and wandering on the red planet and to clear any air of doubt as to the survival capability of any intending colony. A low altitude spot, with less dust as much as possible, close to or on a water ice hidden beneath the surface, close to the equator and as evenly flat as possible, will make a perfect landing spot on the red planet. This cannot be left to guesswork and obviously allays all possible stress after the landing. The possible landing spots finalized upon are but, not exhaustive: Columbia hills [I'm still waiting to see Washington hills, if NASA doesn't mind], Eberswalde, Holden crater, Jezero crater, Nili fossae, Mawrth vallis and others. On these spots, from every analytical view, the first landing colony will find it easiest to get water and power the machineries by solar panel. This is despite having some water taken to the red planet alongside nuclear and other sources as options to powering the machineries.

Then, genetic studies that will completely wipe any atom of worry regarding solar radiation impingement on the red planet are already on-going. This is despite the fact that the spacesuit will give a very good result in protecting man from radiation. It should be noted that the Earth surface is not completely free of radiation but, men and women are going about their businesses, not caring about it. However, the scientific community is ensuring that men and women are shielded on Mars. That's not all; this community of scientists is looking at a supportive approach to enhance the bodily defense of man against solar radiation by a genetic modification technique, so that no one brings the issue of this radiation to mind. This will be looked, a bit differently at again, in Chapter 7.

There's this friend of mine who once said, "I'll love to be on Mars but, it'll be boring there."

So, I laughed very hard. He then looked at me in amazement and asked, "Why the hard laugh."

I asked him, "Mention at least 2 people thinking of these Moon and Mars things we hear on the media." He mentioned them, accurately as I expected.

My next question was, "Do you know their financial status?"

His reply was, "Oh yes! These are the richest guys on Earth."

I then asked this friend, "Will it be a hard thing then, to establish fun and recreational centers by these same guys so that they can equally make some bucks while some people are catching some laughs and tickles?"

He then replied, "I honestly didn't look at it from that perspective. These guys are surely gonna dropped some recreational stuffs."

Apart from recreation, electronic gadgets like phones etc., cars, helicopters and other machineries are already being developed that would work on Mars to ease man's mobility on this planet.

Looking at these above considerations, we can boldly call the mission to Mars a very possible one. This mission is that possible that some persons, youths and adults alike, are already itching to find themselves on the red planet. I'm not just speaking optimistically; everything is figured out and all that's remaining is implementation, after some perfected finishing touches to some gadgets and machineries needed.

CHAPTER TWO

Earth's 'Readiness' for this Move
to Mars and beyond

The heading of this particular Chapter is slightly different from what the conventional expectation might have been but, it's really deliberate. We want to look at a few things and come to that realization that the planet Earth is either 'ready' for releasing some humans to Mars or not. A level of height in technological growth is necessary to be bold enough to spread to other planets. But, should advancement in technology be the only yardstick to start unnecessary adventures of journeying to other planets? Shouldn't we just enjoy our attained apex of technological growth and celebrate, gallivanting in our well-designed aircrafts on the Earth and forget about some dry planets? The truth is this; the journey of the spread of man to other planets and in the future, to other star systems is never prompted by our advanced height in technology alone. Another frank understanding is this; if we get stuck to the Earth and refuse to make efforts to spread, then, this mother Earth will suffer from inability to support us as a species in not very distant future. Trust me; we are not caged to the Earth, I've said it before.

Some well-chosen pointers, apart from our capacity to go on this voyage, will be brought under consideration so that we can fully answer the question, *'Is the Earth ready for humanity's move to become multi-planetary and first, by reaching Mars?'* We want to see our mother Earth better. We truly desire a better environment that's conducive for all. However, shouldn't we know when it's time to extend our numbers to another available space? We should! Since my high school days as a child, I understood when a host wants me to take my leave, even if I've just spent 5 seconds with the person. There are pointers that will make you understand and get the message, though you may be very pleasant, amiable, genial, lovable etc. and when that message comes, which may have stemmed from the host's state or situation, then, you'll have to take your leave. By just looking at our mother Earth, is there any message at all that's telling humanity, "It is time to decongest me, your mother?" Let's see!

According to the United Nations statistics, the world population became 8 billion people on 15th November, 2022 and that's to say the number of people as we speak, exceeds that figure. Some experts have said that the Earth can fully support about 1.5 – 1.8 billion people. That means we are already 'over milking' our mother Earth at the moment, which may cause her to become *cachectic and emaciated* in terms of resources – running down her natural resources in no time – leading to her lack of ability to sustain humanity. If only humanity can spread as-soon-as-possible to other planets, then will our Earth live longer than she would have lived, than if her children refuse to spread. Refusal to spread is actually ending as a 'matricide' i.e. killing our very own mother Earth from the negative effects of overpopulation. So, the Earth is truly saying, "Spread dear children, for my space is becoming too small for you all."

The rate of pollution is on the increase. Be it air, water or land pollution; they are all on the increase. This notable increase is posing greater health-risk than it did over the last decade. Did you know that only 5% of the planet enjoys air that meets international standards for quality at the moment? That's absolutely correct! I live in a town with bursts of industries and companies here and there. It came to my notice after spending just 2 weeks in this town that black soot rests on furniture and electronics in every house that I entered. At first, I thought it was happening just in my house but, after a subtle investigation on my own, I realized that industrial soot is returning back into our houses. If only some persons causing this soot hazard can shut-down their companies and move to Mars, then, this soot and pollution will reduce, right? Yeah! Our

Earth will sure be 'happy' to 'see' that. What about land and water pollution? They are both becoming great concerns already. Bottom line; as people leave for Mars and venture beyond to even live on other planets/moons, less pollution on Earth and more years for this Earth to exist.

Also, greenhouse gases are causing global warming. This situation has started affecting the polar ice caps, melting the ice and increasing the ocean and sea volumes of water. This is obviously the cause of so many observed issues of flooding these days, which will continue as long as we keep growing and remained tied to this planet alone. To help ourselves, some people have to undertake this craved adventure to Mars and beyond, thereby reducing the level of carbonic fuel use on Earth, which eventually reduces the level of greenhouse gases build-up on this planet.

Some scientific claims are beginning to gain ground that humans did not come from Earth. If this claim is true, then, humans will do well in other parts of the universe since we're just visitors doing pretty well in a place initially not our home. That's to say there should be the encouragement to spread and so, decongest our borrowed habitat. Who knows? We may even eventually find our home after beginning the spread, landing first on Mars before continuing the search for our home, if Mars is not our origin either. The spread will truly assist our present mother Earth to 'breath' more owing to reduced number of humans on her surface.

Notably, some diseases and infections are emerging which are now of great concerns to humanity. Major ones among these diseases are communicable and disastrously affect a highly dense population in our societies. Such diseases are known not to thrive in sparsely populated areas and seen to disappear over time in such a scanty population. So, as humanity embarks on the journey to Mars and beyond, the population on Earth will reduce and some major diseases bedeviling humanity will obviously disappear from human population. This will culminate in greater happiness from disease-free states on Earth.

Before this time, forensic, psychological and sociological pointers were showing that there could be a tendency for a 3rd World War, if care is not taken. Wars are devastating to man and to the Earth, likewise. If only some of those fighting for land space can consciously realize that there's abundant land on Mars – for which wars will not be needed to claim and they relocate there – then, Earth will surely be 'thankful' to goodness. This Earth will quickly be 'eager' to 'see' such people leave.

Population growth is at a geometric rate and therefore, the faster the Mars journey becomes a real-time practical possibility, the better. It surely abates all sorts of negativities that can arise from the high population impact. A few of these negativities have been looked at. Glaringly, going to Mars helps those going, to escape from quite a number of things and in turn, helps the Earth by reducing the negative impacts she would have received from those things. Indeed, the Earth is ready to suffer less from the challenges posed by the population explosion; she's truly 'ready' for man's voyage to Mars and beyond, which will ease her stress as some people become permanent Martians.

CHAPTER THREE

Before Deciding on Mars

Before thinking of and arriving at the conclusion that Mars is definitely the easiest candidate among the planets in our solar system, if we are to talk about colonization, another planet was being strongly considered. Venus, the closest planetary neighbor to Earth, was the favorite among the planets being considered by the first generation of modern planetary scientists for colonization and terraforming. Mostly all astronomical observations concluded that going to Venus is shorter than going to any other planet in our solar system, i.e. after taking-off from the Earth's surface and this conclusively placed Venus at the forefront of space exploration research from the very early discovery days. It was even thought and argued by some of the early scientists that it may host life, considering her (Venus) distances from both the Earth and our star. But, these persons realized, though later, that they were dead wrong. So, the question, "What could have wiped the lives on Venus?" arose and prompted greatly ignited interests – even placing her further up on the chart for missions to space to get more facts.

It was known, after the then Soviet Union took the study of Venus further than any other interested group, that she (Venus) had some bizarre characteristics that may not and will not in any way support life in the shortest possible time, unless vigorously altered. [Every world critically-thinking and experimentally-grounded scientist thinks humanity may not have attained what it takes to alter these seen characteristics of Venus yet.] Who will try to go and live on a planet whose surface is a hellish blistering-furnace of an average of 464 degrees Celsius? None! Who'll want to live in an environment where it rains sulphuric acid? Not me and definitely not you, I believe.

About ten probes have successfully and softly landed on Venus, communicated with their guidance computers from the surface after touchdown. Most of these rovers spent over an hour on the surface and sent back data but, they were all destroyed by this volcanically-scotching surface of this planet. Is that where a sane man would want to colonize? No! Not in this generation; neither is it in the near future.

I have seen few write-ups talking about how man can live on a 'hanging city' on Venus and thereby, start the colonization journey from there instead of from Mars. When I heard these none-sense-making write-ups, honestly, my hairs stood on their ends. It's frightening! These fellas said that there's a layer in the atmosphere of Venus that's cool enough for man to live-in but, they've forgotten the sulphuric acid rain part. These funny persons want those colonizing Venus to literally *hang their lives in a balance* because a 'hanging city' can be cut down just as it happens to clothes dried on a rope line. What will then happen to lives after the city accidentally falls to the surface of Venus? Your guess is as good as mine! Fire! One day, I asked one of these guys talking about colonizing Venus, "Who or what will go to the surface of Venus to plant the poles or necessary anchors that the city will be hung-on?" Till now, that question has not received any answer. If it had, in anyway at all received a viable answer, I would have asked this next one, "What will happen to these poles or anchors, even if they got planted and eventually held-up the city, considering the surface temperature and pressure on Venus's surface?" Since the first question had no answer yet, I honestly don't know if he'll find the answer to my intended second question. My point is this; the *hanging-city hypothesis* will definitely not work, not in the near future. It's a very strange and life threatening hypothesis that has been dropped by even the boldest of space explorers known on the planet. Peradventure, when humanity becomes more advanced in technology and attaining a type II civilization status, will a 'flying city' become a possibly attained technological feat. Yes, humanity will get there, if you observe the trend of things. For instance, we're now talking mobile houses and flying cars; there'll be flying houses and rocket cars in the future. Then, humanity will graduate to flying cities and other technological advancements powered by zero-carbon and other fuel sources. Yes, one day, man will technologically 'treat' the planetary sick-state of Venus but, till then, greater discoveries in terms of technical-know-how needed for that will have to be made.

Apart from what has been covered so far in this Chapter, did you know that a day on Venus is longer than one year on Earth? Yes! It's so because Venus rotates very slowly on its axis. I have a friend who doesn't sleep in the daytime, except if he's ill. No matter how sleep-friendly the weather and environment are, he finds it hard to sleep as long as the Sun hasn't gone down. That's just the nature of his body. If this particular guy was to be on Venus, does it mean he'll be awake for more than a year? I doubt if he'll be awake and still sane after a week. Following the extremely long day, an approximately corresponding length of time is taken by darkness on this hot planet. A country is currently almost done with their artificial Sun project; one day, Venus will need such in order to segmentally utilize hours instead of waiting for more than a year before sunrise. That's to equally say research into an equipment or device to induce artificial darkness will be needed in order to possibly break the time on this planet into a 24hr cycle, if man will one day truly spend money and other resources on revitalizing Venus for human habitation after attaining breakthrough technologies to refine her.

Did you know that the atmosphere of Venus smells like rotten eggs? Oh, yes! It is known that this smell is even worse at the level of the sulphuric acid clouds on this particular planet in question. Obviously, living in such an environment – whether on the ground or above it – will be a smelly, non-conducive and stomach-retching experience. This will never be tolerated, not until it's dealt adequately with, and not in this technological era.

The appearance of Venus from the night sky vision is bluish-white and this prompted some persons in the then Soviet Union to think that she may be holding a large quantity of silver. Trying to draw the line of conclusion as to the reality of this possibility, the Union spent over ten probes on Venus alone. After sometimes, it was realized that Venus got her visible hue from galena and bismuthinite; not from silver. Will Mars be richer and more encouraging to humanity than Venus in terms of mineable metals and mineral deposits? Let's wait and see!

From all the facts gathered about Venus – beginning from the onset of this Chapter until now – man realized and concluded that it cannot remain at the top of humanity's space exploration plan anymore, at least for now. Even the Soviet group renowned for championing the Venus exploration, abrogated and forestalled all missions to its surface for now. There are winds of information talking about reigniting Venus exploration in the next decade but, another planet has overtaken her in terms of scientific attention, exploration missions and even colonization. The red planet, Mars, has won the hearts of a greater majority of all the members of the scientific community and he's now on the topmost of the list.

When the scientific community saw the very tough challenges presented by Venus, on taking a look at Mars, it was discovered that though Mars if further away from the Earth than Venus but, he (Mars) presented less challenging attributes. It was realized that he (Mars) has even similar length of daytime compared to what we have on Earth. When talking about surface temperature, pressure and gravity, it would be easier for man to handle them on Mars than on Venus. Mars is only colder than the Earth and that is a far less challenge when compared to trying to handle the hell on the surface of Venus. From these and other drawn facts which we'll see as other Chapters go on, Mars became the favorite planet and tops others as pertaining colonization plans of humanity.

CHAPTER FOUR

Necessary Things on & Around Mars

To be able to live on Mars as a colony, some things are necessary. These things are not just going to be there for any aesthetic reason; they are a must, if some humans leaving the Earth will survive and go about their daily activities on the red planet. These are the needed scientific and technological inventions that will curb hazards and grant people a healthy living on Mars. That is not exhaustive; the daily food, supplies and amenities of all sorts, at least at the basic level are needed until the colony become self-sustaining.

The voyage to Mars is not about just moving a number of men and women to somewhere on the red planet and abandon them to any fate. No! Even criminals will not be treated that way. So, in this case of honorable men and women taking a leap of courage, going on the first, longest, manned spaceship trip in the history of humanity to another planet, those people deserve some restful conditions on touch-down. [Whoever goes on this journey, by the virtue of the courage alone, such a person is honorable.] Therefore, the companies interested in transporting people to the surface of Mars and trying to see that humanity establishes a colony there are not daft; they are working indefatigably around the clock to see that everyone who lands on Mars is as comfortable as possible. I would want to call all such companies **MT1 (Mars Transporters 1)**, made-up of all individuals, private and government agencies in the space race to see that men and women form a self-sustaining colony on Mars. From reliable sources and from research findings, I've realized that some members of MT1 are interested in establishing mining and other companies on the red planet. These members, who are interested in what the planet holds, are in actuality transporting their workers there and are bent on making life as sweet as possible to the individuals who will work for them over there. So, life will not necessarily be at risk as some uninformed persons say. The necessary things are being perfected, so that nobody sues any member of MT1 for negligence of any kind. The loopholes that may be seen by those people going to Mars are being all-round vigorously covered, in order to make Mars as comfortable as a good place on Earth, at least.

Some of the necessary things on and around Mars, as revealed by the various companies planning them are:

Housing
There is a great deal when it comes to housing plan going on for use, when the humans arrive on

Mars. The various members of MT1 have various approaches to the nature of housing that should be inhabited by the human colony. However, these various approaches and designs are all stemming from the vivid considerations made after looking at the Martian environment. The different designs of the supposed human habitats on Mars, if looked at as presented by the different members of MT1, are convergently achieving the same purposes. Some of the purposes are to:

- Shield humans from the radiation reaching the surface of Mars.
- Withstand the Martian wind.
- Be easily heated-up and cooled as quickly as possible, if desired by the occupant.
- Protect the occupant from the dust on this planet.
- Be easily reached by another person who's not wearing a space suit.
- Present the comfort of being able to go unclad and without any clothes-on, as done on Earth, if you so choose.
- Infuse Oxygen into the rooms via a device attached to the housing structure.

Take this; there are members of MT1 who have so much perfected their housing plan by involving various housing companies in a competition. The best housing plan, chosen by one of the members of MT1 does not even require water to be built on the red planet and robots are already programed to see to this particular type of housing on Mars as we speak.

Some persons and countries who are members of MT1 have even gone as far as uttering the names they'll call their Martian cities; that's good. But, that's not really the most important thing. What I've noticeably observed, which is very important, is that everything that pertains to life and comfort are being factored into the Martian cities and that's also really interesting to know.

On Mars, there may be cities: on cliffs, in the ground, inside the Martian rocks and on the soil surface of the red planet. Whichever type a person meets is largely dependent on the member of the intending transporters accompanied to the red planet. The housing and habitats are going to be having automated pressure controls, so that nobody is affected by the low pressure nature of the planet. The design or nature of the habitat does not matter; what matters is a collective availability of those bulleted points above, in every habitat the new Martians will find themselves.

Power Supply

When it comes to power supply, every member of MT1 is leaving no stone unturned. Power supply will determine virtually everything i.e. if not all things. As we speak, the only option not thought of when it comes to power supply is a hydro dam power supply system and that's because the surface of Mars cannot hold water, not yet. [In the future, it would and then, a dam can be constructed.] Apart from that, all forms of power supply are put into cognizance already and special sites are mapped-out on the Martian soil for housing the different forms of energy to be used.

On the list of power sources on the red planet, nuclear source is number one by some members of MT1. These members are saying they cannot toy with human lives and they're therefore going to Mars with this most reliable and long lasting source – nuclear. I evidently realized that though nuclear is no. 1 on the list but, these members said they're equally going to the red planet with very high number of powerful solar panels. I was impressed at this but, apart from the two sources above, high strength solar rechargeable batteries are already designated for use. That is not all. I learnt that some members of the MT1 are seriously researching as we speak, into different sizes of generators that could work effectively in the Martian atmosphere and possibly use different fuel types. When I asked how feasible that is, I was told that if a helicopter called Ingenuity could fly in the Martian atmosphere, then, it's easier for a generator to work.

So, look at the power sources, all going to Mars: nuclear, solar, rechargeable batteries and generators. Should any intending Martian fear any disastrous power glitch? No!

Water Supply

Water will be transported in large quantity, enough for everyone, during the flight to the red planet. On landing on Mars, the water supplied from Earth will still be remaining when water production will commence in each housing unit via the inventions for water production. [About three or even more different types of machines have been invented that can produce water in the Martian environment.]

Some inventors of these machines were contracted to do this by some members of the MT1. However, others just invented them and were noticed, which sparked interests from varying members of the MT1. Without discussing the nitty-gritty of water production process of the invented machines and chemical processes during production – which are both needless in this book – water challenge is already allayed as far as Mars is concerned.

Food Supply
The supply of surplus food is rather the greatest target of all members of the MT1. Excess supply of different types of food is highly needed during the trip to Mars and this supply chain continues until food is successfully produced on the red planet. The primary targets are food types that do not perish easily. However, plans are being made by some members of the MT1 to go with preservation systems for some perishable food to the red planet.

It's been settled that different greenhouses would be established on Mars, in order to grow plants in them and truly end-up with a self-sustaining colony, as planned.

While the plants are still in their infant stage and not yet fruitful, supply still continues to the red planet with food. Food processing plants would be established also, to convert raw materials into different processed foods.

Some MT1 members have even gone as far as ensuring that manual labor is very much reduced and computerized processes would majorly turn out the food needed on Mars. Specifically designed food-producing robots are being looked at and more research is delved into in this area. So that men and women can have time for other things, thereby reducing the human number at the food producing plants and greenhouses to the barest minimum. From what I've seen so far concerning the future of food production on Mars, every Martian would be well fed; I just hope they don't get all fattened-up.

Oxygen Supply

As a matter of fact, this was my first fear for anyone that would be standing on Mars. It's said, a human can do without food and water for days; so food and water weren't my fear for anybody intending to be there, be it an astronaut. But, oxygen cannot be done without for more than few minutes. Well, I have good news for you; oxygen will be very much available in every habitat. Apart from the habitat, well, every spacesuit already has provision for that.

NASA has developed an instrument that's the size of a lunchbox capable of generating breathable oxygen on the red planet. Mars oxygen in-situ resource utilization experiment, or simply called Moxie, has been an oxygen-producing machine from carbon dioxide-rich nature of the planet's atmosphere. NASA's Moxie is seen below.

Again, few members of the MT1 have decide that Moxie's technological make-up may not be just handed unto them from NASA and therefore, they've gone ahead to contract their oxygen-generating machines to scientists who have started recording success in ensuring that the Martian homes, domes, halls, theaters etc. are in the near future, well oxygenated. There is a target in the oxygenation agenda; to have an approximate level of oxygen in the Martian habitats, as seen in noticeably well oxygenated parts of the Earth. This is to prevent any effect that may arise from either over or under-oxygenation of the body.

As a strongly considered point, oxygen-generation is a topmost priority of every member of MT1 and therefore, there are various researches looking into diverse ways to oxygenate the Martian habitats. Some members of the MT1 are thinking of making their oxygenation machines to look like the air conditioners on Earth here. Others are incorporating their oxygenation machine into the water production system. While a few remaining ones are even thinking of having various sizes of oxygenation machines for various spaces of the habitats, i.e. having machines of various space-oxygenating capacities.

Furniture
The future Martians need furniture for the comfort that eventually translates to their health also. Humans will one day be on the red planet but, they aren't going to sit on the sand; neither are they going to stand and sleep. So, furnishing the Martian habitats is very important and some intending members of the Mars Transporters 1 (MT1) have contracted this already to companies to make the lightest of furniture possible. This is because there is need to reduce the pay load on the spaceship so that

other heavier equipment and machines, especially those that pertain to health directly are given greater priority in terms of loading-space on the spaceship.

A furniture company, located in China, makes light and foldable furniture that can be squeezed into small space in the craft. I learnt this company has started talking with one of the members of the MT1. There's a tendency of another member of the MT1 to give the contract of furnishing to the company that will be building the Martian habitat. So, this company will not only build but, equally furnishes the rooms with the rightful and proportionate furniture. In this case, some of the furniture may be permanent structures in the various habitats. Whatever shapes and sizes the furniture come, they'll be comfy.

Spacesuits
From what the members of the MT1 are planning, there'll be no need for spacesuits while inside the city habitats; rooms, walk ways, halls, lounges, theatres, offices, markets, domes, agricultural greenhouses and hospitals will not need any spacesuit. However, going outside the habitats will require this and honestly, this particular aspect has been well covered by the virtue of various improvements being made. Spacesuits that permit greater flexibility than the type used by Apollo 11 astronauts and offering far greater protection from ionizing radiations have been developed. Even handling low atmospheric

pressure, the spacesuit manufacturers have boldly told the space mission markets that their suits can keep humans alive on Mars.

Some members of the MT1 are seriously now looking into digitally-operating spacesuits; the types that can variably alter the warmth, oxygen supply and even issue a warning, if the wearer is approaching a dangerously high radiation environment. It's really going to be sweet to wear one of these types, right? Yeah!

Magnetic Field Source or Radiation-Absorbing Sheet

The human colony on Mars can make do with only a well-designed spacesuit, leading to no need for any of these two in the sub heading above. However, some of the well-caring members of the MT1 are thinking of an additional radiation shielding. This added shielding is going to cover virtually all the surface of Mars and not just the protection given to the human body alone.

Some among the MT1 members are thinking of generating a magnetic field from the poles of the red planet. There's going to be the powering of this field continuously via nuclear energy source. With that, if implementation is rightly done, the amount of ionizing radiation that would reach the surface of Mars would definitely be negligible and as comparable to that reaching the Earth's surface. This would be a very welcomed idea so that the magnetic field surrounding the red planet does not only protect the new Martians from cosmic radiations and charged particles but, also end-up in assisting the initialization of the use of navigation compass on Mars.

A few among the members making-up the MT1 on the other hand, are thinking of using a radiation-absorbing sheet which would be located between the Sun and Mars.

This is slightly tricky but, I've been made to understand that it's possible to achieve. The concept of this radiation-absorbing sheet is illustrated artistically below.

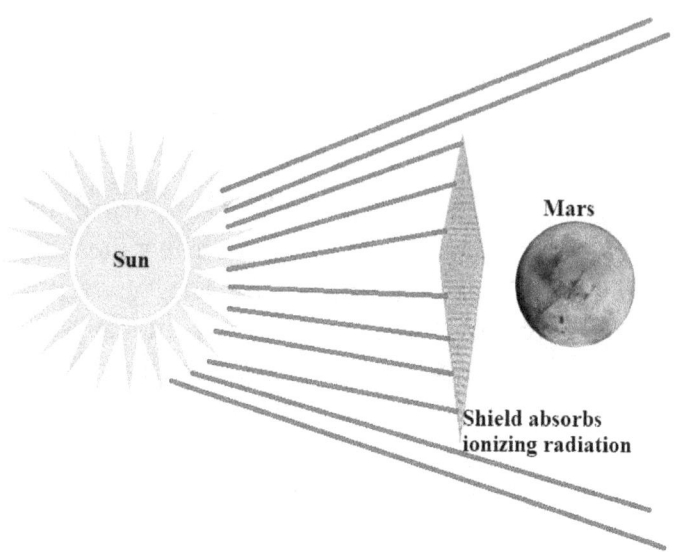

An Engineer said that it (the sheet) would be equipped with light censors that will keep it permanently facing the Sun and backing the red planet. That's not all; it would equally be made to have motion and stabilization capacity so that it can move and remain at a particular distance between the Sun and Mars always. Whichever is implemented in the future depends largely on which member of the MT1 truly prepared for this, because I'm told it requires an appreciable level of finance.

Waste Management/Recycling System

It's confirmed that some members of the MT1 have already designed a waste disposal/recycling system. According to these particular members, it is said that Mars has some debris already on the surface and they do not want more added. So, there's already a penned-down approach to waste management, be it sewage or garbage.

Mars is being colonized with attention to cleanliness in mind and that will not be undermined because some robots, specifically programed for only this, may be brought alongside the human settlers. That is also being looked into, to avoid making a mess that the Martian wind will keep spreading on the surface of the planet.

From the above things looked at, the journey to Mars and living on that red planet is not a suicidal trip. Nobody wants it to be disastrous and it wouldn't be. There are other scientific facts that support the positivity of the mission but, I'll rather not bore you with unnecessary figures and scientific rigmaroles.

CHAPTER FIVE

Robots Go First

Before humans step unto the sands of Mars, robots will go first to prepare the habitats. An idea seen as not only very important but, lays greater likelihood for man to thrive better on the red planet, if the habitats can be made ready before the arrival of men and women. Virtually the entire world thinks it's a workable plan; I think it is, too. It's workable because it involves some of the best Roboticists and Programmers in this world, who are going to see to it that there's shelter for those going to Mars.

Someone I know was concerned about the effectiveness of the robots, if they're sent alone to the red planet. So, I gave this analogy. The twin Voyagers 1&2 are still in contact with the Earth till today, since their programing as far back as1977. Voyager 2, which is closer to the Earth, is approximately 11 **billion** miles from the Earth. On the other hand, Mars, by comparison, is only 40 **million** miles away from the Earth. If Voyager 2 which is billions of miles away can still communicate with the Earth, using an outdated 1977 technology, how much better will the robots communicate using the present day technology from a very short distanced Mars, of just millions of miles away? This person nodded and said he now believes that one of the members of the MT1 is really sure of what he's doing and has hands on deck to achieve this plan. As we were discussing this, a Roboticist called-in and I asked whether it's possible to re-program a robot to do what it was not programmed to do initially. He said, "It's very possible!" and he went on to explain how it's done. Why did I ask that question? I asked that question because things may not really be what they look like on the surface. Yeah, robots are going to Mars first, to set-up the habitats. Should they meet a situation for which they're not configured, does it mean the whole operation will fail? No! They'll be reconfigured from the face of the Earth to handle the new challenge. That's where the issue of communication and the distance falls in. It then means it'll be an easier thing to use the current technology and communicate faster with the working robots on Mars, since it's shorter than the farthest point man has ever communicated with a space probe.

As of January, 2023, three rovers were still in operation on the red planet. The latest of these 3, Perseverance, has done over earthly 2 years after landing on the red planet on 18th February, 2021. While the biggest Mars rover ever made, Curiosity, has spent over 10 years and still in operation. Why am I bringing this up? This is because one of the members of MT1 said his company needs just 2 earth years to build the habitats that would sufficiently contain at least, a thousand settlers on the red planet. That's to say, with the improvement in technology before these habitats-building robots get to Mars and spring into action, the work should be a success and if rovers made earlier are active for over 10 years, then, these robots which will build the Martian homes should do better in terms of longevity. With that said, the feasibility is sealed with a stamp of positivity; it's possible to attain robot-built Martian habitats. As a matter of fact, that's the best way to go.

When we take a look at the members of the MT1, many of them are intensifying the production of different types of robots. Some are contracting these machines out to Roboticists to assist in designing and production. A few others are concealing their plans to utilize robots on the red planet. But, if only a vivid look is taken at the present trend of robots boom, it would be realized that diverse robots of numerous capabilities are being built today. Some of them, if not a majority, are really being designed for use on Mars.

Some skeptics may still not be convinced that the going first and setting-up of the habitats will involve robots. But, it's already a done deal. Some members of the MT1 have even signed agreements with Roboticists and their companies; that should their robots land successfully on Mars, a specified length of time should be enough to produce a city that can hold a particular number of people. Roboticists for some of the MT1 members are eager to see their works in fruition on the sands of Mars already, because they not only boast of those robots that can build only but, robots that can fix other robots without any guidance from the face of the Earth. That's not all; I learnt that some robots will be fixing satellites without taking permission from anyone to do it. Some humanoid robots may be site managers right on the red planet in this work to come and to help in seeing quickly to the success of the plan, they are given greater level of intelligence.

Now, the level of success in landing, recorded with robots on board, will either make or mar the confidence of those humans intending to go and live on this iron-filled planet, in the members of the MT1. Everyone who is thinking of Mars as an option will be waiting breathtakingly to see how it went with the spaceship that took the robots. The men and women who are ceaselessly seeing going to Mars as a home-coming will hold their breath as the spaceship descends to the surface of Mars with robots. It will be a moment that will either convince them or deter them from embarking on this journey. A wise space transporter must just get this right; he cannot just blow this chance to get people rallying to his company and leaving for Mars.

Robots, by the virtue of their material make-up will end-up weighing more than humans and therefore, to land successfully on Mars with them shows the great extent of dedication given to preparation. When the journey begins – while lifting off the surface of the ground with robots on board – the spaceship motion can be technically analyzed, so that experts end-up revealing whether there's a gracefully easy take off or not. Since escaping Earth's gravity into orbit is one of the toughest stages during missions, the ease with which the spaceship is able to beat the gravity, having stacked robots will imbue a lot of people with confidence in the member of the MT1. Then, a steady, non-wobbling, upward and bulleted acceleration from the ground, if possible with robots, tells the entire world that such a transporter will do far better and record even greater acceleration when only humans will be on board. To have a safe touchdown with robots, tells the magnitude of the perfection which the spaceship has in terms of balance and maneuverability, even while descending through the thin atmosphere of Mars. The world is going to trust more the handiwork and creativity of any member of the MT1, if there's a hitch-free landing with robots on Mars. This would be the first impression on the red planet and therefor, a safe landing translates automatically into a greater success through patronage of that member of MT1 who records it.

According to an analyst who's also interested in being on Mars, he explained that if any member of the MT1 should load robots from the Earth and successfully off-loads them on Mars, it means the control from the guidance computer on Earth is top notch. He said it doesn't stop there; it equally means there's appropriate tacking of robots and loads in such a manner that the center of gravity of the spaceship is not compromised in anyway. He concluded that with these two things in place and always observed, the touchdown on the red planet will always be a smooth one. He added, "Whoever among the members of these intending Mars transporters gets the landing with the robots right, that's the spaceship I'll accompany to Mars."

In less than a decade, they'll be touchdowns with robots on Mars. All eyes are on the different members of the MT1 to compare and contrast who's the best, just judging from the first landings on Mars. If Mars is your destination, please, go with the best.

CHAPTER SIX

The Spaceships

The spaceships for the Mars missions are the life-carriers to move men and women from the face of the Earth to Mars. These crafts are going to be holding at least a hundred people for approximately 7 months on transits. That's not all; some spaceships, depending on the design or the transporting company, may have cargo on board in addition to the humans on this journey. When it comes to these ships, the transporter either gets it right or else, a disaster is unavoidable. However, all intending transporters are fighting tooth and nail to ensure there's no disaster at all during the voyage to Mars, by putting a perfected touch to every designed spaceship to undertake this mission. Moreover, the journey is a lengthy one and therefore, comfort, after assurance of safety, is the next target by the members making up the MT1.

Some spaceships to the red planet may be all-purpose type, where humans, cargo and even animals (yes, some members of MT1 are transporting animals, too) may be on board at once. In this situation there's a greater deal of compartmentalization and partitioning of the interior of the spaceship to hold all these at once and it involves a level of technicality, if all must fit in well. But, a few of the members of MT1 are thinking differently. They are thinking of having the spaceships for humans differently made from the ones for cargo, which will equally be both different from the tanker spaceship in low Earth orbit. This group of transporters is trying to avoid the complex dexterity in partitioning the interior of the spaceships, avoid the mixture of cargo with humans and see to it that the humans-only spaceships are lightened to travel faster to the red planet. These few intending transporters are thinking of:

1. **Spaceships for Transporting Humans:** they'll be designed with great emphasis placed on interior compartmentalization. [As said earlier, some members of the MT1 are designing all-encompassing spaceships.] These ships to transport just humans and their food, clothes and drinks, will be given a detailed and admirable interior partitioning. It's the target of the few members of the MT1 thinking of this, to make this particular spaceships as comfortable as possible to all humans on board. This particular ship will be the state-of-the-art ship among all the spaceships for this mission. The transporters thinking of this approach are saying that it will be clumsy and unmanageably difficult to

lump-up humans, cargo, robots, food and animals in one spaceship. So, the specially designed spaceships for conveying humans are going to be different only in the internal comfort architecture from other ones; not in engine or technical capacity in anyway. There's going to be detailed planning for fun-filled flight to Mars. We're talking about almost 7 months here; virtually all members on board must be happy and in good mood during this time. If not, one unhappy fellow can cause trouble for all and the spaceship in flight. Care is taken to have as much entertainment on board as possible during this voyage. From what I heard, an inside source whom I wrote to, said that those on board may not know when the 7 months will whisk-by during the voyage to Mars, as all will be enjoying themselves on this trip. I was made to know that the partitioning of this spaceship for humans will include the following but, not exhaustive: a sleeping cabin, wash/restrooms, kitchen, dining, restaurant/lounge, hall, gymnasium, laundry, swimming pool, board games room, computer games arcade, a clinic and others. Beautiful, right? Yeah! But, apart from the comfort targeted by the members of the MT1 trying to adopt this approach, it will have a positive effect on the travel time, by shortening it. This is how it happens; there's going to be a reduction of the overall load carried by the craft and not carrying these particular spaceships' full load capacity – something deliberately planned. With the reduced load on board, the thrust from raptors and engines will give a resultant increased acceleration and velocity during the voyage. Thereby, reaching the red planet faster than the time which the cargo ship will take. It's an approach adopted to lighten-up the spaceship for faster trip and easier maneuverability, generally. However, these particular spaceships with humans on board will be taking-off last, so that others carrying cargoes will touchdown and get the habitats set with furniture and other things, before the arrival of humans.

2. **Cargo Transporting Spaceships:** these are going to be, according to some members of the MT1, transporting only cargoes and crew members on board. Not all intending transporters are thinking of this but, I think it's worthy of consideration. The MT1 members interested in cargo transporting spaceships are saying that they've successfully cut down cost of compartmentalization, as these ships do not require exquisitely demarcated segments on board. Rather, what they try to achieve is good cargo stabilization by fastening them in place, so that there's a smooth movement during flight. There'll be entertainment and fun centers on these particular ships but, their sizes, capacities and variations will be reduced when compared to the spaceships to transport humans. On these, the use of the total carrying capacity will be the order and these ships go after the robots have set the habitats in place on Mars.

3. **The Refueling Tanker Spaceship:** this will be a spaceship that's having about 99% of its volume designated to carrying fuel into low-earth orbit. It then awaits the spaceship proceeding to Mars, which has escaped the Earth's gravity; refuel this spaceship to full tank before it continues the journey. It's very important for this provision, as escaping the gravity of our planet here involves an appreciable use of rocket fuel. Refueling has to be done when a spaceship reaches low-earth orbit so that there's assurance of making it to Mars; as half-tank rocket fuel may not be enough to propel the ship to Mars. Not all members of the MT1 have spoken about this spaceship but, they're very much aware of its importance and the dangers of trying to boycott it from the Mars mission plans.

Orbital Flights and the Federal Aviation Administration

It will be a good idea, if the Federal Aviation Administration (FAA) of the United States considers encouraging orbital flights and not in any way discourage them via stern criteria. This is because there is a space race on the Earth now as we speak and whenever I mention the MT1, I'm referring to all countries, government bodies, private companies and individuals with the intention of transporting men and women to the surface of Mars. The race for Mars is an obvious one and there are eyes already on the possible treasures on the red planet. It's equally glaring that some countries are even training the Martians-to-be already; I saw a documentary.

The FAA has to fast-track the orbital flights, encourage all private and government space industries in the US – who are saying they are ready for this flight stage – to get ready and those ready should be scheduled. This is because it will be a slap, if the US is beaten in this race to Mars by another country, especially by a country that's treacherous. Is that all? Let's look at! Could any country in this space race lay claim to Mars, if the US is beaten? Maybe we should just ponder on that question.

[Please, if you're a citizen of another country and not US, don't take this personal. Why I wrote this is because a few countries, I won't mention names, are thinking of saying Mars is theirs, if they ever set feet on it before any other country. Only the US is ahead to beat these countries and so prevent anyone from laying claim to Mars. As it stands, only the US can stop that planned nonsense from those few countries and so, Mars will become a habitation for all nations.]

So, it's imperative the FAA ensures US is not delayed in this race for the red planet. The conditions to be certified for orbital tests can be reviewed and made easier. Rules are made for man and not the other way around. It will be honorable for the great US not to end-up taking permissions before landing on Mars or even worse, told by another country that there's no place for the US to land on the red planet. Let's consider it wisely and ensure the US makes it to Mars first, so that the entire world and human colonies on Mars can enjoy peace.

CHAPTER SEVEN

Sustaining the Martian Colony

There's a great yearning on the hearts of few people that I have met, who want to go and live on Mars, permanently. The yearning for doubled assurance that the Martian colony will truly be self-sustaining is upon their minds. These people have questions with many 'What if' starting them. One of them; what if they run-out of food? Pointing at supply and the possibility of sustenance, right? Yes! And, yes!! The colony on the red planet will be self-sustaining with time, not immediately. The truth is that every member of the MT1 that's already talking about a time frame when humans will be on Mars, have factored this aspect (sustenance) into the grand plan of colonization. The Martian colony has to be sustained, if not, going to the red planet is a suicide, futile and baseless mission.

To take the colony on Mars to a self-sustaining and healthy level, the following have been planned by some members of the MT1:

1.) *Taking only law-abiding people to Mars.* This is the 1st great step to sustaining the new

Martian colony. A few of the MT1 members have decided that for Mars to be sustained, it starts with having only law-abiding persons on it. They took this decision and defended it rightly from this angle; a criminal, especially an ex-convicted burglar, could go and steal something from the powerhouse on the red planet, if s/he is a pathological burglar and thereby cause problem for the entire population. So, this is a very good point to consider in selecting those going to Mars. Yes, those going have to be crime-free, by all records. However, I realized that some members of the MT1 have not included this step to sustenance of the colony on Mars into their plan – they should do that for the sake safeguarding lives. There are speculations that even if everybody is, by all records free of crime, any criminal act conclusively traceable to anyone on the red planet, will lead to eventual return of that nefarious person to the Earth. This is to ensure the red planet doesn't quickly become a den of criminals.

2.) *Going to Mars with only sane humans.* Some of the other MT1 members are thinking of ensuring that all persons going to the red planet are in good state of mental stability. This particular perspective is also an addition to the list of criteria by the few intending transporters who hold no.1 above as another point to consider.

It's another way to have lives and property safeguarded. To this effect, everyone going to Mars will pass through a sanity test; a test that must be passed, to be allowed to go and live with other men and women over there. I'm very much in support of this because it may be disastrous if a mad man or woman gets there.

3.) *Productive greenhouse farming developed.* The different members of MT1 are seriously looking into this, because this is going to be the Martian colony's source of food on the red planet, after the supply from the Earth is cut off. [But, as long as the greenhouse farming system has not yielded enough to take care of the Martian colony, supply spaceships will be coming to deliver food items.] Modalities are put in place to ensure that effective production, within the shortest possible time, is obtainable from the greenhouses that will be located on the red planet. Looking at the importance and necessity of this – not taking it granted and neither handling it with levity – some of the members of MT1 have contracted out this project to ensure that there are computerized operations in the production processes, so that human errors are eliminated and the greatest of produce is attained from this project to sustain humans on the red planet.

4.) *Animals are coming with humans to Mars.* It's intriguing to know that some animal species will be found on this planet. Yes, a few members of the MT1 have decided that some animals will reach Mars; other members think that animals will pollute the red planet quickly. Some of these animals will be keeping humans company (companion animals); others will be few selected food animals in growing pens and the rest, experimental animals. This has been decided by a few intending transporters who are interested in ensuring that Mars has a supply of meat and other animals, as the future colony to be found there becomes self-sustaining. Indeed, if Mars cannot produce meat and meat products, then, the autonomy wouldn't be complete. So, these few members of MT1 are thinking really deeply.

5.) *Health research laboratories will be looking at how mammalian body reacts to and how to quickly adapt it to the Martian environment.* Some of the animals reaching the red planet will be delineated to research purposes. These animals are going to be mammals like us, so that general studies that can be extrapolated to humans can be conducted on them. Majorly, changes brought about by the new environment upon the body systems will be studied, some animals may be exposed to the environment and findings will help humanity know how to adjust and live better on the planet, Mars. The research is not intending to use humans as experimental guinea pigs and therefore, experiments as to how man can quickly adapt, will come from these intending laboratories which would be seen on the red planet. The more man adapts, the better the sustenance because the more robust will the health of everyone be.

6.) Clinic or hospital to attend to health issues. The red planet has no living disease-causing pathogen. It is disease-free as we speak. However, there'll be at least a clinic or better yet, a hospital to attend to health issues and challenges. This is in a bid to close all loopholes and looking critically at it, a health challenge could be from another angle, not necessary a pathogen-caused illness. There could be a fracture, etc. and so, it'll help in seeing to the health-improving state of the population on Mars.

7.) The Tardigrade gene experiment and modified humans, genetically. A few of the members of MT1 are thinking of staging a special research center on Mars. This center, which may be bearing fruition, may genetically modify the permanent settlers on Mars, to withstand the radiation on this red planet. From research findings, experts have come to believe that the DNA of Tardigrades would prove useful to the new Martians since these animals can survive the vacuum of space, zero temperature and withstand deadly ionizing radiation levels. If this is done and success achieved, then humanity would be said to have conquered the radiation on Mars and thereby, needing no fear of it anymore. A Tardigrade is seen below.

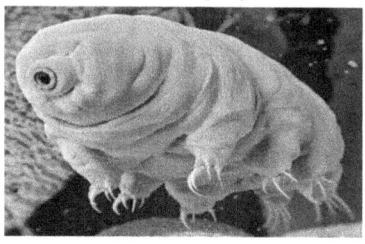

8.) *Technological and other research laboratories would be found on Mars.* Not only health researches would be conducted on Mars. Technological researches would commence right on this planet, with the aim of seeing how to mine and utilize the available resources on this planet to better the lives of men and women on it. It doesn't end there; some companies targeting this planet for gain would try to see how they can make a financial fortune by mining some resources, return them to the Earth and sell in either raw or processed states. Also, after getting to Mars, humanity may confidently choose to advance the interplanetary trip to other parts of the solar system. It's then known that a record has been set with Mars and adding a little to that, even from Mars, can take humanity further away from the Sun. If such a trip is to be made from the Earth, it would be longer. But, from Mars, it would be shorter to reach places further and so, the red planet becomes the cradle of further humanity's missions, if technological laboratories are set-up on Mars to assist in colony's sustenance and deep-space research advancement.

CHAPTER EIGHT

Terraforming Mars

Terraforming (in simple literal sense, 'Earth-shaping') of a planet, moon or other cosmic bodies to end-up having Earth-like environment, atmosphere and become habitable, can be done to the red planet. Yes, scientists from the various fields that concern this are looking deeply into it. There are so many proposed theories out there as to how to achieve terraforming; theories from which diverse and voluminous series can be written on *Terraforming Mars*, if the various ways possibly are to be considered in details. The concept of achieving a terraformed Mars is basically hinged on:

- Building-up the magnetosphere.
- Raising the temperature.
- Rightful atmospheric content and pressure build-up.

These three major changes above can be achieved faster by some drastic techniques but, man would not be able to be found on the planet during this time. However, there are other techniques which can be deployed, though slower but, humanity can be found on the red planet while these other techniques take their turn and course. It is not that humanity cannot live and move on Mars continually without having it terraformed. A portion of mankind that would have seen moving in pressurized suit on the red planet as fun would have still inhabited it, even if terraforming is never mentioned. Terraforming and the necessary steps and scientific rigmaroles involved may not really be the immediate headache of mankind, which speedy and hasty procedures should be experimented with. It requires perfection to an extent and the more the preparation, the better the outcome.

Humanity has gotten to that stage that Mars is a project that cannot be abandoned anymore. So, diverse theories as to how to terraform Mars are projecting out of different scholarly bodies. Some of the ways and theories being put forward are so bizarre and unrealistic with our current technological growth on Earth. Some individuals even say that the Earth doesn't have the technology for terraforming Mars; as though this requires a particular machine that will be switched-on with a button. They're dead wrong! Terraforming is not about machines as they think. Others feel it will be overly expensive to achieve a terraformed Mars. Some of the other theories being proposed are not overly expensive. They are just observed procedures, majorly from industrial techniques since some persons and governments are interested in having industries on the Martian soil. The only thing required of these other cheaper techniques is consistency. With consistency, the Martian atmosphere and temperature will have the breathable capacity needed. That's what some members of the MT1are looking into; having a cost-effective company and/or industry which produces something on Mars, while the activities from this company and/or industry reshapes the atmosphere of the red planet, bringing habitability to Mars again. I think this makes perfect sense, if it can be attained. So that it wouldn't be that money is spent on trying to terraform Mars alone but, equally generating capital from this procedure. Great!

Man will be on Mars within a decade from now. Whether the process of terraforming starts or not before man sets foot on the red planet, it is neither late nor will it become impossible to attain. As a matter of fact, the terraforming process that will take place while man is on Mars is the best because parameters can be measured easily, with accurate precision. Humanity will breathe oxygen on Mars in about 200 years' time; that's not a very long time from now.

I do not want to project a particular terraforming technique as proposed by any member of the MT1, for obvious reasons. However, these intending Mars transporters have their various terraforming plans in the tunnel. Some members of the MT1 think that it is better the terraforming start while nobody is on the planet yet. Others have a contrary opinion. However, scientific analysis going-on as relates the various ways and proposed theories for terraforming shows that about 50%, not all, are making sense and feasible with current technology. That is to say; indeed some members of the MT1 will get it right and Mars will one day have a breathable atmosphere.

It may not be about nuking Mars poles; it may not even be lasers or mirrors from space, etc. Simpler ways of achieving this without spending much are being considered and they are making sense. Introduction of genetically modified and better photosynthetic organisms, the melting of polar ice caps and greenhouse gases release, can be useful in attaining a warm and oxygen-rich atmosphere that will be breathable. Then, to ensure the atmosphere is retained, the magnetic field of the red planet has to be kick-started and various workable ways are in the minds of the MT1 members. The planet's lack of a protective magnetic field, if not reversed, will keep stripping the positively attained changes from any terraforming program off. It's therefore of a great importance to even fix the issue of the magnetic field first, if speedy results are to be attained from some particular techniques that will involve greenhouse gases. As we speak, there are experts trying to look into the possibility of having a production company that will still meet the positive terraforming goals mentioned in this particular paragraph.

Some scientists have come-up to lay it bare that if Mars is allowed to have various terraforming techniques going-on at about the same time, from different members of the MT1, it may result in having jumbled beehive of confusing activities on the red planet. According to these persons, if different techniques to terraform Mars are going-on at the same time, there may be a counter effect from procedures that are not synchronized to achieve a uniform result or a specifically targeted outcome that should be worked fully towards by all. Therefore, these people are suggesting and it makes sense to me, that the terraforming technique to be used should be agreed upon and strictly adhered to. So that in doing such, all outcomes from the various members of the MT1 to be found on the red planet can cumulatively achieve the desired effect faster without any counter force in the established protocols. In this situation, it is the desire to work in unity among all those intending to transport humans to Mars that will prevail. That is not all; all possible ways of achieving terraforming have to be looked-at and only one will be picked and instituted. Some individuals or private companies that are members of MT1 must realize that some of the techniques they are having in mind must be laid bare before the entire MT1 members in this case – it's for the good of the entire mankind – and not to conceal any technique in terraforming procedure. To lay bare, means Mars will end-up being terraformed faster; to conceal means there could be delay in achieving this desired outcome, especially if there's a simple and cost-effective way but, hidden from others. Never should it be considered a technological giveaway to lay terraforming plans bare by any member of MT1 because man has surpassed the stage of being selfish to his own detriment. It would be better to consider all and then, pick one method. When Mars is terraformed, the resultant effect is what is gotten from an artist's impression as 'B' below.

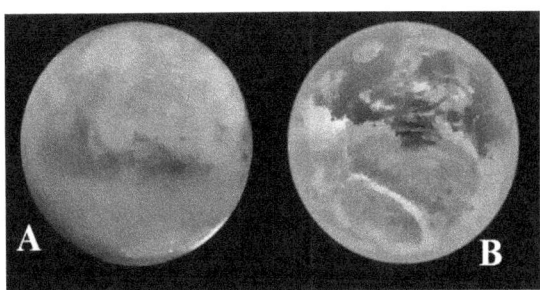

A: Before Terraforming   B: After Terraforming

Approximately 10 billion USD was spent on the James Webb Space Telescope, the greatest invention so far in the field of astronomy and kudos to NASA for getting it right. Did you know that some terraforming techniques on Mars could cost less than that telescope? Yes, as long as the terraforming is done by a revenue generating invention on Mars. So, government agencies among the MT1 members like NASA and ESA have to be ready to solicit funds from all government bodies on Earth, who are interested in a place on the red planet. With that, if success is attained, which I really expect, then, Mars will be terraformed faster than the time-length we are projecting.

CHAPTER NINE

The Prospects on Mars

The red planet has myriads of opportunities already awaiting the new human colonies. Some may laugh this off but, not so fast. Wait! A planet with fortunes spread on the surface, a place where people are racing to in order to hit an unearthly record, a pathogen-free planet, etc. has attracted thousands of people already. So many people: professionals, skilled, semi-skilled and unskilled men and women are already warming up for Mars, even before I decided to pen the first word in this book. Remember that the target of the members of the MT1 thinking of getting humans to Mars is to end-up having a self-sustaining colony; the red planet will need diverse disciplines to reach that self-sustenance. You reading this, if you want have a change of your cosmic location, fine! On Mars, there'll be a place or even places for you to fit-in. If you think you have to become a Martian, trust me, you reason is genuine.

Did you know that some members of the MT1 are going to fly some people to Mars for free, if these people can't foot their fare? Yes, you saw rightly, 'for free.' I heard from an insider, that some people would have free flight to the red planet. Those to enjoy this privilege will be working on the red planet for these particular members of MT1. Before boarding the spaceship, an agreement would be signed to the effect of having the person as a worker on Mars and when s/he begins to receive his/her monthly earnings, in bits, the fare to Mars will be deducted from his/her monthly earnings over a period of months. It is the plan of this member of the MT1 to have not only the most courageous and mentally stable people to embark on this journey but, to go with the most useful, even if they do not have the footing capacity to their fares to Mars. From this and other plans by other members of the MT1 – who are equally thinking of a way to strategically subsidize the transportation fare to Mars, so that they can have the best hands possible – the red planet will be having busy men and women from virtually all disciplines, if they want to be there. I do not know specifically the exact plan for subsidizing the fare to Mars from other MT1 members but, they are sure planning something. So, if you are thinking of being on Mars permanently, it may not be as expensive as you think, especially if you are going there to work for any member of the MT1 that will transport you there. This particular paragraph has revealed that it's not about transportation anymore; it is about making-up your mind for this change, a positive one for that matter. Are you ready? Let's see!

On Mars, from the look of things and depending on the winner of the race from the MT1 members, the currency that's most likely going to be used on this red planet is the dollar. [Maybe it will be called the Martian Colony Dollar (MCD) for differentiation sake.] However, it may actually result in having various currencies on the red planet, if the leading racer to Mars delays and other MT1 members from different countries make it there almost about the same time. Again, if some MT1 members have companies and industries in mind, their workers will be paid based on the currency of their country of origin. Why am I going this far? This is because it's better to bring it mind now that from the MT1 members racing for Mars, if you're to consider going, you'll be paid in one of the topmost currencies of this world.

I did mention that there are many things to engage one, gainfully, on the red planet. A few of them and hints to others will be seen as we go on. A building engineer that I met online is already gearing-up for this planet. He told me he's going to the red plant under the auspices of any MT1 member that's willing to pay him on Mars. Why I'm mentioning him is this; he told me categorically that he'll work for any member of MT1 and he'll look at what seems like ruined ancient civilizations on Mars. He said, "Using my engineering knowledge, if I truly draw the conclusion that there were civilizations on Mars, I'll equally dive into Archeology to present my findings to the world from a YouTube channel." Can you see the opportunities he sees? Let's list them; Engineering, Archeology and YouTube. Yes, three! This engineer has just one job on Earth now as we speak and he's not comfortable with his remuneration from it. But, he's thinking of bettering himself, right? Yeah! A few of NASA's photos of Mars, which this engineer saw and thinks Mars may have held a civilization are seen below.

I've heard some people say, "Mars is so dry; there's nothing there. Oh, it will be boring." They are wrong! If only they're thinking like this engineer friend of mine, sighting the fortunes and opportunities on the red planet, they'll have known better. Kindly open your mind's eye and see better, if you're thinking of an alternative to this Earth.

I've met a young man who loves following what the great Blue Origin and notable SpaceX are doing when it comes to rocket engineering and preparing for space missions. He's a nutritionist. My conversation with him is seen below.

One day, he said, "I love what Jeff Bezos and Elon Musk are doing, when it comes to their space and rocket industry." He continued, "I would want to be on Mars one day but, I doubt if Mars would be exciting."

All I did is this; I asked, "What are your hobbies?"

[He made mention of his hobbies and I took mental note of them.]

From the hobbies of this young man, I showed him two things he could do on the red planet that would fetch him additional income, even if he attaches to any member of the MT1 and works as a nutritionist or not. As we speak, his target is to be on the first flight for permanent settlers heading to Mars.

Honestly, personally, if I'm to be on Mars – knowing fully well that Earth's atmosphere is thicker than Mars's and remembering that the likes of Galileo discovered planets using primitive telescopes – I would have started preparing for Mars with a telescope, even if I don't know jack about Astronomy. The very thin atmosphere of Mars would pave greater way for clearer vision of the telescope. Who knows what you could discover with a better telescope than Galileo used, while peering into the universe from Mars? Something could come-up, you know.

Sometimes back, when I came across Olympus Mons, the tallest mount in the solar system, I was excited. Standing at about 25km high, it is approximate two and half times Mount Everest's height.

I actually got excited when I saw this and wished I'm a mountaineer but, I'm not. I'm not dabbling into mountain climbing in the future but, I know that one day, the tallest mountain in the entire solar system will have its peak reached by a brave Martian or a number of them. That record may be awaiting you, if this is your area of specialty. However, if this is not your area, look at Mars again; there's something it could offer you, higher than what the Earth may offer you.

In geographically structured contrast to the above paragraph, the largest canyon in the solar system, Valles Marineris, is also located on Mars. It extends over 3,000 km long, 600km across and 8km in depth. Valles Marineris is seen below.

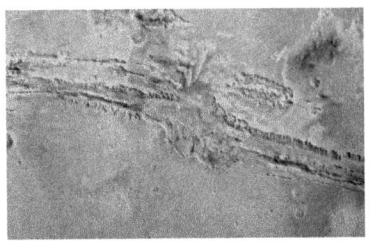

If you're a structural engineer, an architect or a builder, what can you positively do with this? I leave the canyon to those who have studied what it takes to transform it to something else, after landing on Mars. That's how to see some of the fortunes held by Mars.

In a group chat on a social media, a young man said he's leaving the Earth for Mars because his immune system is found to be medically weak towards some diseases-causing organisms. I actually sympathized with him and I told him that he's thinking of doing the right thing. I was made to understand that because of this predicament of his, employers of labor aren't happy with his health whenever they employ him. So, with time, they just fire him. According to how he sees it, which is correct and which makes this scenario fits here, on Mars, his body wouldn't battle pathogens and he wouldn't have to be sacked because of pathogenically caused illnesses anymore. Mars will thereby present a better working prospect, in terms of freedom from diseases, than the Earth for him.

No one has to be a professional before being able

to have the ticket to Mars. Mars will need all people from every occupation, labor, profession and job ever known in this present day on Earth. You've gotten few hints and ideas about what the red planet presents. These cannot be exhaustive; kindly extrapolate them to your own area of life, discipline, hobbies, etc. If you're interested in being on Mars, look at it again; there's something in stock for you there, something others are not seeing.

CHAPTER TEN
The Future of any Martian Colony

It is the plan of any member of the MT1 to ensure that any group of humans that lands on the red planet is well-catered for, until a self-sustaining state is attained. Before the autonomy, the sustenance of the colony of humans will be seen-to by supplies coming from the Earth. These supplies would include things not readily available yet on the red planet, but processes to have them in the future would have kick-started by then. [I deliberately reiterated this again, so that nobody would think s/he would be brought to Mars to perish.] Additional people would keep landing on Mars, until about a million people would be found here. It could be more, because people would leave all walks of life to be on Mars. However, from what the different members of the MT1 are planning individually, the colony on Mars may attain self-sustenance before the one million population mark is attained.

The Martian population will grow but, that growth, availability of basic amenities and food supply cannot guarantee the continual existence and progress of the Martian colony. The new inhabitants of this planet from the Earth must observe continually, the necessary safety rules needed to support life on this planet. Every system has a set of rules guarding and guiding it and the red planet will not be an exception when man finally stands on it. Obedience from every member of the colony, to every set rule and regulation is a must. Remember that in Chapter 7, we looked at the necessary things for establishment of self-sustenance and now, we're considering the advancement into brighter future after the sustenance is attained. There are similar but, there's a line of demarcation. From that Chapter 7, it was seen that law-abiding people should be taken to Mars. Now, it's being said that only continual observance of law and order, unbroken obedience on the part of every colonist on Mars to set rules and regulation will brighten the future of the entire Martian colony.

The red planet may have held a civilization in the long past; some scientific evidences are pointing at such possibility and a boy born into Russian family, Boriska Kipriyanovich, revealed that he reincarnated from Mars. This story started making headlines since 2004 and experts, psychologists and philosophers have looked critically into it. Yes, Boriska was unusually fast in development after his birth and his knowledge about Egypt, Mars and some other planets was correctively on point. He even mentioned some things that humanity is yet to fathom-out and earthly sages on the plain of reincarnation believe he's telling the absolute truth. Now, Boriska said Mars was destroyed as a result of nuclear war and that he has come to Earth to warn humans not tread on that path. Why am I saying this here? This is because Mars will most likely become an experimental ground after humanity stands on it. There will definitely be health-related researches to be conducted in view of bettering the life of man on the red planet – a greatly welcomed idea. But, it would be better and safe to the lives on the red planet, if nuclear experiments are not part of the experimental designs planned by any of the members of MT1. [By this, I'm not referring to nuclear power plant for electricity generation.] From interviews with Boriska, the Martians were far ahead of the humans in technological terms. If it's so and if they could not control their anger during a strife, which resulted in nuclear war – then it's better to keep nuclear experiments away from the red planet – since this planet will be holding different groups of human colony, from different nations, who may be brought to Mars by different members of the MT1.

For the brighter future of the human colony that will be on Mars, brotherliness and unity are strongly advocated, irrespective of the origin, religion, culture, tradition and other life attributes of each and every one of the inhabitants-to-be. To bond and unite will help well so that information can quickly circulate, i.e. should some not get any information at the point of dissemination among the targeted population. The unity and bond will not only help to attain peaceful coexistence alone; it'll equally stimulate quick springing into action of mostly all at once, in order to attain a set colony goal. This unity and bond will equally help the social status of everybody and the ego of the entire human population on the red planet, if these two things are both present. Trust me on this; segregation, separation, detachment, avoidance and other social errors will wreck the population with time, by starting slowly to affect moods of some individuals in the colony. The negative effects on moods could aggravate into mental disorders in the population. From this notable point, it'll pay better if Mars will have higher number of extroverted cheerful men and women. It'll be better you fall into that happy category, if you're preparing for Mars but, not yet such a person. A positive change can start, gradually, from now.

One the red planet, everyone must be listened to and anybody who sees what must come to the attention of others, especially a technical issue, must speak. Why am I raising this point? This is because I've seen foolishness, stupidity and self-destroying pomposity in a few groups of people on this Earth. I've seen where a man wanted to genuinely pass information that will benefit everyone in a group but, he was shut-down by the leaders in that group. When the day came that his warning should have averted the feared repercussion – if only he was listened to – the repercussion met everybody except him. He was exonerated because the authorities overheard that this man wanted to warn all but, nobody listened. There's another aspect to this; saying what ought to be said but, keeping mute. Let's look at this; there was an apartment building that had electrical issues and all the tenants knew it. Sparks of electric current were seen from the connection point of the block to the main electric line but, everybody kept quiet over it. The day it happened, all the houses on that block got burnt; none was spared. This happened because the people refused to speak-up, despite seeing what was looming over them. So, to listen to anybody that has something to say and to say anything worth noticing should be the way to live progressively on Mars.

In my life, I've seen instances where a group of people was made-up of very sane, law-abiding and lovingly-compassionate people and suddenly, there's a change in someone. A change that defies what the group stands for and contradicts the expected attributes from this person. The person suddenly becomes the opposite of what s/he used to be. In such a group, if positive results should still come forth, then, the negatively changed person must be dropped.

As it concerns Mars, obedient, sane and positively interactive people are expected to be heading there. If on getting there, someone becomes moody, schizophrenic, abusive, and tending towards any noticeable insanity, such a person must be returned quickly to the Earth. The person can be returned via the supply spaceships return trip back to Earth from Mars, to avert any danger from such a person because such is more dangerous on Mars than on Earth.

Also, it's advisable everyone who'll be on Mars is engaged doing one thing or the other. With that, every mind will be engaged, positively. This can advance the planet faster than if some busybodies are very negatively busy doing nothing good. If people are just left without any engagement on the red planet, their minds could be usurped by negative emotions and thoughts that could end-up being disastrous to some inhabitants or even the entire colony. With that said, I think the members of the MT1 have to decide from the face of this Earth where anyone going to Mars could possibly fit-in before taking such an individual to the red planet.

Mars must be free of all forms of stigmatization, in order to advance faster as a colony, in the future. Talking about stigmatization, there are diverse forms of it and it'll pay the future Martian colony if none is found. Let me sight this example; there's a lady warming-up for Mars now but, she was born as a baby boy. A lady but, born as a boy? Yes! Don't get it twisted; she never for one day went for transition hormone therapy. I met her online and she said, "People stigmatize me thinking I went to change my sex through hormone therapy." She continued, "I just started noticing those changes when I became 14 years old; I've never thought of doing that myself." She went through counseling to be able to stand the societal stigma. But, bottom line, she's leaving for Mars since she's still facing stigmatization. I told her to try and advocate for recognition of four (4) sexes on Mars, to completely put an end to this stigma thing over there, when she lands on the red planet. You see; Mars must not recognize two (2) sexes just as Earth does. The mistakes of Earth shouldn't be replicated on Mars. I told this lady that even if I'll remain on Earth, I could assist her to advocate for the recognition of the 4 sexes on Mars, to prevent stigmatization of any form. These four (4) sexes are: M, F, TS and HM. Their full meanings, just to clear any cloud of doubt are:

M – Male

F – Female

TS – Transexual

HM – Hermaphrodite

CHAPTER ELEVEN

Conclusion & the Forecast Welfare
of the Intending Martians

Within the limits of the current technological framework of the world, colonization of Mars is possible and it's a non-life threatening reality. It is not just about going to the red planet and surviving as a colony alone; it is about having an autonomously existing independent population of humans on another planet. It doesn't just stop there; all loopholes leading to life threatening issues must be closed and researching into advancing humanity's existence to further planets and habitable moons will vigorously be embarked upon, as success with Mars infuses man with greater confidence and poise to pursue further distances in the solar system. The journey will not even stop in the solar system. It starts with Mars, then, other parts of our solar system and finally, man becomes stellar in about 2,000 years from now.

When talking about closing loopholes to all life threats, alternatives and other options to so many things, equipment, ways, etc. have been fathomed and would be deployed, so that the Martian colony is never stranded in anyway. There's always, at least, another way to do anything in the universe. Living on Mars is another way man will live to demonstrate our adaptability to our own solar system before we start talking about another star system, in the far future. Demonstrating this adaptability does not stop with Mars; it actually starts with Mars. It starts with Mars so that when our Sun begins to expand and engulf planets in the far future, man would have gone stellar by then. Reaching Mars is not the final goal in humanity's coverage of the cosmos; getting to Mars, is just the first step, and this step must be taken with accuracy and preparedness. There are greater challenges than Mars ahead and humanity must conquer all. If the red planet cannot be conquered, we're stuck to the Earth and it's not designed to be so.

Companies, some already awarded contracts to handle different aspects of life on Mars, have researched and are still researching deeply into ways to make the red planet habitable and self-sustaining. From reliable sources, these companies are given specific aspects such as: housing, feeding, air-conditioning/pressurization, health, etc. These are given singular and specific responsibility; they are not bombarded with whole conglomerated all-life facets, to avoid cumbersomeness. They are designated in a specific life area, so that every loophole to failure can be covered. From what I see companies and individuals doing; researching, brainstorming, producing, developing, etc. lives and properties on Mars will be well-guarded. I already foresee safety on the red planet that I'm tempted to be jealous of those who have made-up their minds to live permanently on this planet of concern.

The twin Voyager 1&2 spacecraft were launched by the National Aeronautics and Space Administration (NASA) in 1977 and they are still in operation. They have both gone far out of the solar system, after over 45 years in flight. Did you know that both spacecraft have memory under 70KB each? Yes, 70 kilobytes is a small amount of memory space today but, those twin spacecraft are working till now. Let's take it from this perspective; 70KB data space could take a spacecraft to far beyond our solar system. Greatly improved data spaces are available today, even up to Terabytes (TB). Shouldn't it be far easier to land people on shorter trip to Mars with the highly advanced computers of today? It should! If the twin Voyager 1&2 are actively functioning till today, using an outdated technology on board, then, Mars landing and colony existence can be better seen to by the advanced computers of the present time.

I hate being a pessimist and I don't necessarily express baseless optimisms. Martian colony self-sustenance is already done and dusted, only awaiting vivid manifestation in the near future. The Earth definitely will give some humans to Mars and if you're one of those humans, don't be afraid because everything is well mapped-out already. If you may not be interested in the red planet but, you have someone interested, please, don't discourage the person and neither do you need to be anxious over this person. I was already moved by the story and enthusiasm of a brave young lady, Alyssa Carson, who's just in her early 20's. This highly optimistic lady wants to be on Mars. She's not crazy; she's only rightly looking in the direction which elders before her ought to have looked. Since the elders have lost the courage to move, a youngster has courageously gotten-up to this task. I have equally seen other young people who are already gearing-up for this red planet. They are not mad, either. They're the most jovial, amiable, strong, social people I've ever met; they're like warriors in their happy state after a victory. Really, these are the people Mars needs; they are the type that'll awake the sleeping planet and make him smile again. I have personally met one of these young people interested in Mars, who told me that he's elated as though thinking of going to Mars is a returning home thing. I wasn't surprised to hear that. Why? This is because some people on planet Earth are actually Mars's seeds and the story of Boriska (the boy from Mars) buttresses this. The particular young man I was saying I met is an engineer now and he's quite brilliant. He looks muscular but, he never engaged in body building before. He doesn't remember anything relating to past life on Mars but, he feels very good about going to live permanently on the Martian soil. I had this conversation below with him, when he raised the discussion about going to Mars with me, seeking my opinion about his yearning to be on Mars.

He said, "Doc, I really feel like being on Mars and not returning back to this Earth." Then, he asked me, "Do you think there's something wrong with me?"

I could see the honesty in his eyes and from every psychological look at him, there was absolutely nothing wrong with him. So, I replied, "No! There's nothing wrong with you." But then, I went on to ask this, "Is there a specific thing or person you want to run from? Could that be your driving force to be interested in Mars?"

He shrugged his shoulders and said, "I am at peace with people. Nothing is really the Earthly reason for my wanting to go to Mars." He went on to say, "I just wish there was an Earthly reason."

When I asked about his childhood, I learnt that he had a very fast development in all facets of life, faster in all ramifications than other normal children. I realized that I could have truly met a Mars seed. So, I told him to gear-up and start putting his mind together for approximately 7 months trip.

He now asked, "I heard there's no gravity in space and I'll have to start gymming to retain my muscles. I don't like that stress."

That's when I realized that his muscles may be natural but, I wanted to be sure and so, I asked, "Are you saying these muscles are natural?"

He replied, "Yeah! I've never been into a gym before."

I truly saw a naturally built body that could be used profitably. So, asked, "Do you know Mike Tyson?"

"Of course, the 'baddest man on the planet' is what he's called," was the reply of this young man.

So, I asked him, "Did you know that the intending human transporters to Mars must equip their spaceship with a sort of gym on board?"

His reply was, "Yes! I think I've seen something like that on the ISS, where some NASA astronauts were exercising." [ISS stands for International Space Station]

I started smiling broadly and when I saw his anticipation, I jokingly said, "You can strengthen those muscles on your 7 months trip to Mars and then, become the 'baddest guy on Mars' after touch-down, you know."

We both laughed hard and I said, "Well, it's possible!"

He said, "I love boxing but, I don't have the reflexes for it."

[Let's stop at that. We actually discussed few other things.]

Who'll become the 'baddest man on Mars' in the

very shoes of the great Mike Tyson? We're waiting to see it. Mars will never be boring as some people think. Myriads of sports will be seen on the red planet, starting with indoor games. Are you a sports person? Even you have a place on the red planet. Do you intend to be on Mars? With all these Chapters put together, I think the future Martians may enjoy a better welfare than Earthlings, if the members of the MT1 will execute their plans to an appreciable excellence. Then, know this; whatever you do, whoever you're and wherever you're from, the beautifully red planet, Mars, will welcome you with an embrace.